Mom Duck led a trip. Quack, quack! Quack, quack!

Step off the curb and then cross. Be quick, so you will not get hurt.

It is Quinn's turn to cross. Be quick, Quinn!

Do not turn to the left! Turn back, Quinn!

Quinn is lost. Burt the Cat can help him.

Burt and Quinn step off the curb.

Quack, purr! No one is hurt!
Quack, quack, purr!

Surf is up. Jump in quick.

But do not get hurt!

The End